Make a Mind-Controlled Arduino Robot

Tero Karvinen and Kimmo Karvinen

O'REILLY®

Beijing · Cambridge · Farnham · Köln · Sebastopol · Tokyo

Make a Mind-Controlled Arduino Robot

by Tero Karvinen and Kimmo Karvinen

Published by O'Reilly Media, Inc., 1005 Gravenstein Highway North, Sebastopol, CA 95472.

O'Reilly books may be purchased for educational, business, or sales promotional use. Online editions are also available for most titles (*http://my.safaribooksonline.com*). For more information, contact our corporate/institutional sales department: (800) 998-9938 or *corporate@oreilly.com*.

Editor: Brian Jepson
Production Editor: Teresa Elsey
Technical Editor: Ville Valtokari
Cover Designer: Mark Paglietti
Interior Designers: Ron Bilodeau and Edie Freedman
Illustrators: Tero Karvinen and Kimmo Karvinen

December 2011: First Edition.

Revision History for the First Edition:
    December 13, 2011   First release
See *http://oreilly.com/catalog/errata.csp?isbn=9781449311544* for release details.

ISBN: 978-1-449-31154-4
[LSI]
1323797733

Contents

Preface

Shortly, you will build your own mind-controlled robot. But that's just the beginning of what you'll be able to do. As you follow the explanations for components and codes, you will thoroughly understand how your robot works. You can keep applying the knowledge to your own robots and EEG-based prototypes.

You'll learn to

- Connect an inexpensive EEG device to Arduino
- Build a robot platform on wheels
- Calculate a percentage value from a potentiometer reading
- Mix colors with an RGB LED
- Play tones with a piezo speaker
- Write a program to avoid lines (tracks)
- Create simple movement routines

From Helsinki to San Francisco

In spring 2011, O'Reilly invited us to Maker Faire, which is the biggest DIY festival in the world. We had to come up with a gimmick for the festival. It had to be both new and simple enough so that everyone could understand how it worked. Cheap EEG devices had just arrived to market and we thought that it would be interesting to couple one of those with a robot.

As a result, we demonstrated the first prototype of the mind-controlled robot at Maker Faire. It was a hit. People queued to try controlling the bot after seeing it in action, as you can see in Figure P-1.

The bot is easy to use. You put on a headband and when you concentrate, the bot moves. Focus more and it goes faster. And it's a real robot too; it avoids edges so that it stays on the table.

We built the first prototype (Figure P-2) with Ville Valtokari. The robot part was based on soccer bot from *Make: Arduino Bots and Gadgets* (O'Reilly, 2011). We read the EEG with a NeuroSky MindWave. The early model had to use a computer as a gateway between Arduino and MindWave, because we

Figure P-1. *Attendees enjoying our robot at Maker Faire 2011, San Francisco Bay area.*

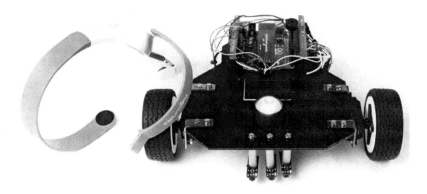

Figure P-2. *First prototype of the Mind Controlled Robot.*

were running the MindWave software and our own Python program on the computer.

Maker Faire was great. Arduino was clearly the platform of choice for hardware hackers. There were Arduino robots that could dive and others that could fly. So did we stand a chance of getting any attention to our little bot?

Reactions

"It's a fake!" Our favorite reaction was disbelief, as it showed that EEG tricks were still new. As if what we were doing was so amazing that it simply had to be just a magic trick. We only heard this about five times, though.

Most of the users simply thought the project was cool. Some were a little skeptical at first, but trying is believing. About 300 visitors tried the device and many more were watching (see Figure P-3 and Figure P-4).

Figure P-3. *Robot at Maker Faire 2011, San Francisco Bay area.*

We were surprised that it could work in a setting like that. Our prototype could handle hundreds of visitors. Also, the NeuroSky EEG headband was easy to put on and didn't need any user training.

A couple of visitors had probably played with EEG before. They just noted "Yep, it's a NeuroSky" and started talking about something else. Luckily, Brian Jepson had made a 3D-printed version of the soccer bot, so we had a backup gadget to amuse them.

EEG in Your Living Room

Control a computer with just your mind. On one hand, it sounds almost like a sci-fi fantasy. On the other, EEG (electroencephalography) was first used in the early 20th century. What kept you waiting for the future?

EEG is the recording of electrical activity of the brain from the scalp, produced by neurons firing in the brain. The brain cortex produces tiny electrical

Figure P-4. *Attendees control our robot at Maker Faire.*

voltages (1–100 μV on the scalp). EEG doesn't read your thoughts, but it can tell your general state. For example, EEG can show if you are paying attention or meditating.

The tiny voltages are easily masked by electrical noise from muscles and ambient sources. EEG currents are measured in microvolts (μV), which are millionths of a volt:

```
1 μV = 0.001 mV = 10⁻⁶ V
```

Noise from muscle and eye movement can be quite powerful compared to this. In normal buildings, the electrical main's current radiates a 50Hz or 60Hz electromagnetic field. In a laboratory setting, EEG is usually measured in a room that has less interference. At home, the EEG unit must filter out the troublesome signals.

EEG devices used to be prohibitively expensive and annoying to connect, and the data required expert knowledge to interpret. For many years, a starting price for the cheapest EEG units was thousands of dollars. They required conductive gel to connect. Having very clean hair and skin was recommended. Most units used at least 19 electrodes. EEG results were printed on paper and doctors had to take a course to be able to analyze them.

Now EEGs are cheap, starting from $100 (USD). Devices are available in shops and by mail order. Consumer-level EEG units are manufactured by NeuroSky and Emotiv. (OCZ used to make a similar device.) With the Open-EEG project, you can even build an EEG device yourself.

NeuroSky's units are the cheapest option, starting from $100 for the Mind-Wave (shown in Figure P-5). The headband is fast to attach and works on dry skin without any gels. It only needs electrical contact on your forehead and earlobe. NeuroSky devices measure attention and meditation as well as the raw brainwave data.

Figure P-5. *The robot in action.*

Emotiv EPOC promises to recognize multiple visualized thoughts. At $300, it's not very expensive. The Emotive EPOC headset also measures head tilt and muscle activity.

OCZ used to make the mOCZ Neural Impulse Actuator (shown in Figure P-6), which cost about $100. It made multiple measurements, mostly concentrating on muscle activity.

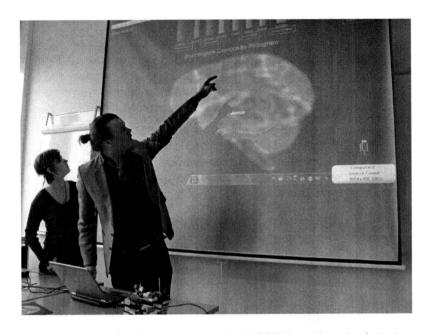

Figure P-6. *Haaga-Helia student project with OCZ Neural Impulse Actuator.*

The OpenEEG project provides instructions on how to build your own EEG device. It's the most expensive option, costing about $400. Building the device requires both technical skills and understanding of safety issues. After all, EEG involves connecting wires on different sides of your head!

NeuroSky MindWave

In the mind-controlled robot presented by this book, a NeuroSky MindWave is used to measure attention. For about $100, you get a CD-ROM, a headband, and a USB dongle for the wireless connection.

The headset has a single electrode with a ground and reference. This means that there are two metallic things touching your head. The measuring electrode goes on the left side of your forehead. In the EEG lingo, this point is called Fp1. It's F for frontal, followed by p1 (1, the first odd number, indicates 10% to the left of your nose; 2, the first even number, indicates 10% to the right of your nose). The other electrode, reference point, goes to your left ear (A1). The headset measures the voltage between these two electrodes.

The CD-ROM that comes with the MindWave contains software to be used with the headband. It can show you attention level, meditation level, and connection quality. You first place the electrode on your head and adjust until the connection is good (the poorSignal value will be shown as 0).

When you focus on something, your attention level (0–100%) goes up. You can do some math, read something, or just concentrate on your fingertip. When you relax, your meditation level goes up. For example, you can close your eyes and take deep breaths. If you can calm and focus your mind at the same time, both attention and meditation can go up to 100%.

The USB receiver dongle can be hacked to connect directly to Arduino.

What Do You Need to Know?

This is a book about building a mind-controlled robot. It's not meant as the first book on beginning with Arduino.

If you are just getting started and want a beginner book on Arduino, see our *Make: Arduino Bots and Gadgets* (MABG) from O'Reilly (2011). We'll point out relevant chapters below.

Before you start, you should have experience in the following:

- Basics of Arduino
  - Installing the Arduino IDE
  - Running Arduino's most basic program (Blink)
  - Writing your own simple Arduino programs
- Basic mechanical building skills
  - Soldering
  - Drilling

You need to know the basics of programming Arduino. You should make sure you can run a "hello world" or "blink" example on your Arduino before you try anything else. This means that you should also have the Arduino IDE installed. If you need help with this, see "Starting with Arduino" (page 18) in MABG. For hand-holding walkthrough code examples, see any of the projects in that book. You may also want to look at Massimo Banzi's *Getting Started with Arduino* (O'Reilly, 2011) if you need a beginner's introduction. However, as a prospective robot builder, you will find the projects in MABG an excellent complement to the one in this book.

Running Ubuntu Linux? Starting from Ubuntu 11.04, you can *sudo apt-get install arduino*. For other distributions of Linux, see *http://www.ardu ino.cc/playground/Learning/Linux*.

You should have basic mechanical building skills. You'll solder wires to an RGB LED and your own connections to MindWave. To build the robot platform, you'll need to drill some holes. For soldering, see MABG, "Soldering Basics" (page 47). For drilling, see MABG, "Building a Frame for the Robot" (page 217).

Acknowledgments

We would like to thank:

- Haaga-Helia
- Tansy Brook, NeuroSky
- Valtteri Karvinen
- Nina Korhonen
- Marianna Väre

Conventions Used in This Book

The following typographical conventions are used in this book:

Italic
  Indicates new terms, URLs, email addresses, filenames, and file extensions.

`Constant width`
  Used for program listings, as well as within paragraphs to refer to program elements such as variable or function names, databases, data types, environment variables, statements, and keywords.

`Constant width bold`
  Shows commands or other text that should be typed literally by the user.

`Constant width italic`
  Shows text that should be replaced with user-supplied values or by values determined by context.

TIP: This icon signifies a tip, suggestion, or general note.

 CAUTION: This icon indicates a warning or caution.

Using Code Examples

This book is here to help you get your job done. In general, you may use the code in this book in your programs and documentation. You do not need to contact us for permission unless you're reproducing a significant portion of the code. For example, writing a program that uses several chunks of code from this book does not require permission. Selling or distributing a CD-ROM of examples from O'Reilly books does require permission. Answering a question by citing this book and quoting example code does not require permission. Incorporating a significant amount of example code from this book into your product's documentation does require permission.

We appreciate, but do not require, attribution. An attribution usually includes the title, author, publisher, and ISBN. For example: "*Make a Mind-Controlled Arduino Robot* by Tero Karvinen and Kimmo Karvinen (O'Reilly). Copyright 2012 Tero Karvinen and Kimmo Karvinen, 978-1-4493-1154-4."

If you feel your use of code examples falls outside fair use or the permission given above, feel free to contact us at *permissions@oreilly.com*.

Safari® Books Online

 Safari Books Online is an on-demand digital library that lets you easily search over 7,500 technology and creative reference books and videos to find the answers you need quickly.

With a subscription, you can read any page and watch any video from our library online. Read books on your cell phone and mobile devices. Access new titles before they are available for print, and get exclusive access to manuscripts in development and post feedback for the authors. Copy and paste code samples, organize your favorites, download chapters, bookmark key sections, create notes, print out pages, and benefit from tons of other time-saving features.

O'Reilly Media has uploaded this book to the Safari Books Online service. To have full digital access to this book and others on similar topics from O'Reilly and other publishers, sign up for free at *http://my.safaribooksonline.com*.

How to Contact Us

Please address comments and questions concerning this book to the publisher:

> O'Reilly Media, Inc.
> 1005 Gravenstein Highway North
> Sebastopol, CA 95472
> 800-998-9938 (in the United States or Canada)
> 707-829-0515 (international or local)
> 707-829-0104 (fax)

We have a web page for this book, where we list errata, examples, and any additional information. You can access this page at:

> *http://shop.oreilly.com/product/0636920021780.do*

To comment or ask technical questions about this book, send email to:

> *bookquestions@oreilly.com*

For more information about our books, courses, conferences, and news, see our website at *http://www.oreilly.com*.

Find us on Facebook: *http://facebook.com/oreilly*

Follow us on Twitter: *http://twitter.com/oreillymedia*

Watch us on YouTube: *http://www.youtube.com/oreillymedia*

1/Building the Chassis

Before programming anything, we'll build the chassis for the robot. Basically it's a traditional rover robot structure with two servo motors in the front and one caster in the back. To make it suitable for mind-controlling needs, we'll add a line detector and RGB LED on the top. We use a solderless breadboard and the ScrewShield for the Arduino, to make adding components and wires easy. Figure 1-1 shows the design of the chassis.

Here's how all the major components will work together to create a working robot:

Arduino
> This is the brains of the project. It is essentially a small embedded computer with a brain (a microcontroller), as well as header pins that can connect to inputs (sensors) and outputs (actuators).

Chassis
> This holds everything together. It's essentially the platform for the robot.

Servo Motors
> These are motors that can be connected directly to the Arduino without the need for any additional hardware (such as a motorshield). The Arduino communicates with them by sending pulses to control speed and direction.

Caster wheel
> Because we'll be turning the robot by varying the speed and direction of the servos, which are fixed in place, we need one wheel that pivots nicely. A furniture caster is perfect for this, and the robot ends up being able to rotate in place.

RGB LED
> This component changes color and tells you what is happening in the code, so you don't have to divide your attention between the serial monitor and the robot. It also gives instant feedback for the users when they try to move the robot by focusing.

Line Detector
> With the line detector, your robot will avoid a black line, which makes it stay in the arena (helpful for keeping it from falling off a table).

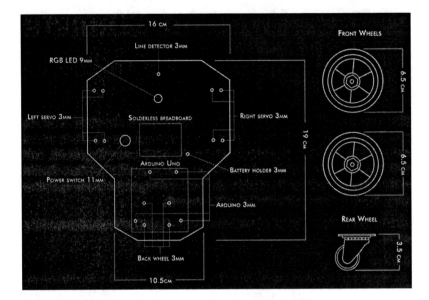

Figure 1-1. *Blueprint of the chassis*

ScrewShield

> ScrewShield adds "wings" with terminal blocks to both sides of Arduino. Terminal blocks have screws, so you can attach one or more wires firmly to any pin. This prevents wires from popping out, which makes building and coding the robot much easier.

MindWave

> MindWave measures your brainwaves and transmits the results for the Arduino. We have to hack the MindWave dongle a little bit because we want to connect it directly to Arduino instead of a computer's USB port.

Tools and Parts

Here we list the parts and tools needed to make the robot. Feel free to improvise if you don't find the exact matches.

Parts

Figure 1-2 shows all the parts you need for this project.

1. Base material (we used Dibond)

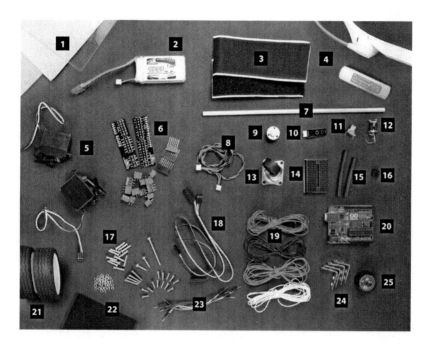

Figure 1-2. *Parts*

2. Rechargeable battery (we used a DualSky 1300 mAh 30C 2s1p 7.4 V battery)

3. Velcro

4. NeuroSky MindWave

5. Continuous rotation servos (we used Parallax [Futaba] Continuous Rotation Servo) (2)

6. ScrewShield

7. Aluminum pipe, 8 mm thick, at least 60 mm long

8. Connection wire for the line-detecting sensor

9. RGB LED, common anode

10. Line-detecting sensor (we used DFRobot's Line Tracking Sensor for Arduino)

11. Potentiometer (rotary, linear resistance); choose one that can be easily inserted into a breadboard (we used one with maximum resistance of about 10 kOhm)

12. Power switch (any two-state switch will do)

13. Furniture wheel (caster) with ball bearings

14. Small solderless breadboard
15. 4 mm heat shrink tubing
16. Piezo speaker (we used one with Rated Voltage [Square Wave] 5Vp-p, Operating Voltage 1-20Vp-p)
17. Screws: 3x20 mm (4), 3x10 mm (13), 3x16 mm (12), 3x42 mm (1), 3x18 mm (2); nuts: 3mm (30).
18. Servo extension cable (2)
19. Ribbon cable or assorted wire in four different colors
20. Arduino Uno
21. Wheels (such as Parallax part number 28109)
22. Felt pad
23. Jumper wires
24. L-brackets, with 3 mm holes and 1 cm between the holes, to match the holes in the servo (4)
25. Reflector for the LED (optional)

Tools

Figure 1-3 shows all the tools you need for this project.

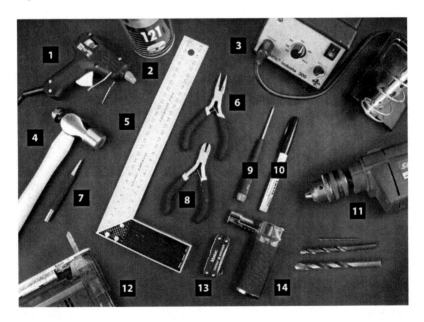

Figure 1-3. *Tools*

1. Hot glue gun
2. Spray paint
3. Soldering iron and solder
4. Hammer
5. Try square
6. Pliers
7. Nail punch
8. Diagonal cutter pliers
9. Phillips screwdriver
10. Marker
11. Drill (3mm, 9mm, 11mm bits)
12. Jigsaw or metal saw
13. Leatherman (wire stripper, flat screwdriver, small blade)
14. Torch or lighter

Servo Motors

Servo motors (Figure 1-4) will be moving the wheels of our robot. The most usual type of servos have limited rotation. They are used when you need to turn the motor to a specific angle. In our robot, we only need to control speed and direction. And, of course, the motor needs to be able to turn freely. Continuous rotation servos are made for this. Almost any servo can be modified to continuous rotation, but it's easier to buy a ready-made version.

The Parallax (Futaba) continuous rotation servo is perfect for our needs. It has an external potentiometer adjustment screw, which allows identical centering of two servos effortlessly. You'll notice how handy this is later, when we program the movements for the robot.

 If you want to learn how to modify any servo to continuous rotation, read the Soccer Robot chapter in our book *Make: Arduino Bots and Gadgets* (MABG), published by O'Reilly (2011).

Attaching Servos

We're going to use regular L-brackets to attach the servos. Attach two brackets to each servo with 3x10 mm screws, as shown in Figure 1-5.

Figure 1-4. *Continuous rotation servos*

Figure 1-5. *L-brackets attached to servo*

If you can't find suitable L-brackets, you can make them from metal strips. For example, you could salvage some strips from an old typewriter, drill holes that match your servo, and bend them to a 90° angle in the middle.

Chassis

For the chassis you'll need something that is robust enough to hold the robot together and can be shaped easily. Plywood, acrylic, or metal plate (Figure 1-6) works well.

Figure 1-6. *Plywood, Dibond, acrylic*

Our material of choice is Dibond, which is aluminum composite with polyethylene core. It's light, easy to cut, strong, and good looking. Best of all it happened to be free. Because it's lightweight and flat, it's used for printing advertising signs. Lucky for us, printing doesn't always go like it should, and tons of Dibond ends up in the trash.

Even if you don't have a sign-making store as your neighbor, you could find other useful material thrown away. For example, you shouldn't have any trouble finding a metal plate from computer parts or plywood from furniture. Choose a material that is easily available for you and that is comfortable to process with your tools and skills.

You don't need to limit yourself to traditional methods when making the chassis. If you have the access and necessary skill for 3D printer or laser cutter, go for it. Go to *http://www.thingiverse.com* for some inspiration on printed 3D objects.

Let's start by drawing the shape of the robot (Figure 1-7). Yours doesn't have to be exactly like ours. Just make sure that the wide end is at least 16 cm and the narrow end at least 10.5 cm wide. Our bot's overall length is 19 cm.

Cut the shape out with a jigsaw or regular metal saw (Figure 1-7). You can sand the edges after cutting to remove possible sharp corners and ugly cutting marks.

Always use hearing protectors and safety glasses when using power tools. Safety glasses should be used when you are cutting, bending, drilling, or soldering.

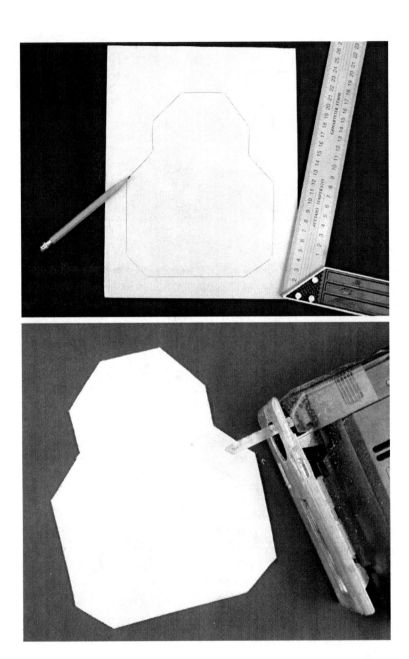

Figure 1-7. *Outline of the robot and the shape of the chassis cutout*

Making Holes

The easiest way to get the holes where you want them is to hold the target object as a stencil and mark spots with a pen (Figure 1-8).

First, mark the servo bracket places on both sides of the robot. The back wheel will be mounted centered in the back of the chassis and the line-detecting sensor centered in the front. Place the Arduino in the back of the robot so that the screw holes don't overlap the back wheels' holes. One more small hole is needed to hold the Velcro for the battery. Place it so that there is space left for the solderless breadboard.

Every hole we marked so far will be drilled with a 3 mm drill bit. We still need a couple of bigger ones: a 9 mm hole for the RGB LED and an 11 mm hole for the power switch.

Next we'll drill the holes, but first, it's hammer time. Always use a nail punch and a hammer to make a small starting hole before drilling metal (Figure 1-9).

Painting the Chassis

We used high gloss black spray to paint our bot. Spray multiple thin layers to achieve a sleek and durable coating (Figure 1-10). You can paint both sides of the chassis. We left the bottom unpainted so that it would easier to see which side is which on the photos.

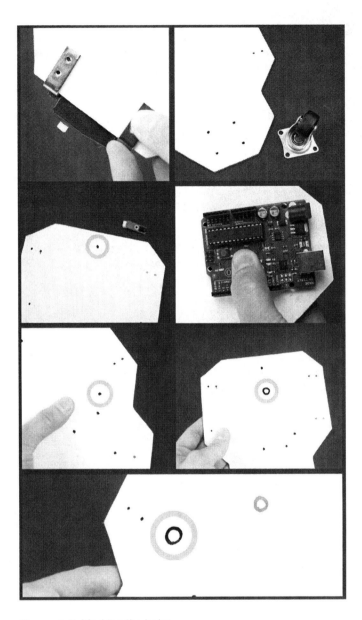

Figure 1-8. *Marking the holes*

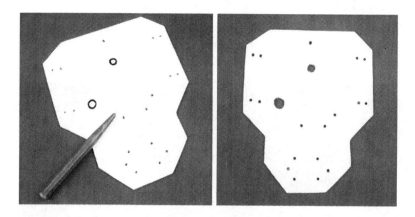

Figure 1-9. *Small starting holes made with a nail punch, and holes drilled*

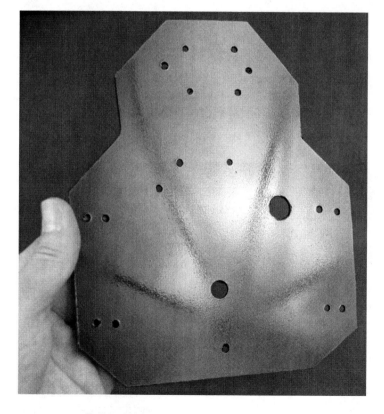

Figure 1-10. *Painted chassis*

Attaching Servos to the Chassis

Next, attach the servos to the chassis with 3x16 mm screws (Figure 1-11). Place them so that the wires are pointing forward.

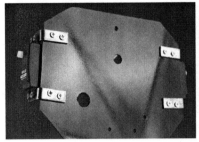
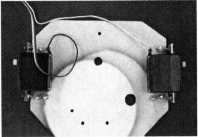

Figure 1-11. *Servos attached*

Attaching the Line-Detecting Sensor

Cut a 3 cm piece from the aluminum pipe. To make sure that the pipe does not short-circuit the sensor, we'll put a round piece of felt pad on the other end. Make a hole in the felt pad so that you can push the screw easily through it (Figure 1-12).

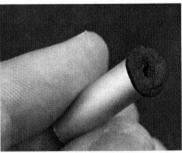

Figure 1-12. *3 cm piece of aluminum pipe for the line-detecting sensor and a felt pad to prevent short circuits*

Put a 3x42 mm screw through the pipe and secure the line detector in the front of the robot (Figure 1-13). The emitter/receiver part of the sensor has to be facing down.

Figure 1-13. *Line-detecting sensor attached*

How does the line-detecting sensor work? It has a infrared emitter and a infrared detector. Reflective surfaces bounce the infrared light back to the infrared detector. So we know there is no line. This does not happen with nonreflective surface, such as a black line. Generally reflective surfaces are white and nonreflective ones are black. Just keep in mind that this is not always the case. We have had black paper that was more reflective than our white tape. With three or more line-detecting sensors, you can make a line-*following* robot. Instead of just turning around like our robot, line followers try to keep the black line in the center sensor. If the side sensors detect the line, the robot will turn until the center sensor sees the line again.

Wheels

There's plenty of choice for the wheels (Figure 1-14). Just don't pick any that are too heavy, and make sure that you'll be able to make the holes for the servo horn attachment. For our robot, we salvaged tires from a remote-controlled car.

Drill two 3 mm holes through both servo horns and the wheel rim. Secure the servo horns to the wheel rims with 3x10 mm screws. Push the servo horns on the servo shaft and tighten them in place with 3x18 mm woodscrews (Figure 1-15).

Figure 1-14. *Selection of wheels*

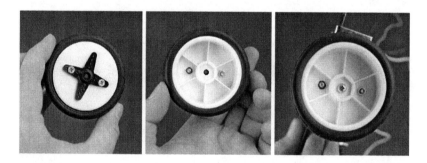

Figure 1-15. *Servo horn attached to the tire rim and wheel attached to the servo*

Attaching the RGB LED to Chassis

As the name implies, RGB LEDs are able to emit red, green, and blue light or any combination of these. Later you'll use the RGB LED to show the level of attention you measure with the EEG headband.

Put hot glue around the bottom of the RGB LED hole we made earlier. Stick the RGB LED in the hole from below and keep it still until the glue sets (Figure 1-16). Be careful to not touch the hot glue, as it can cause burns.

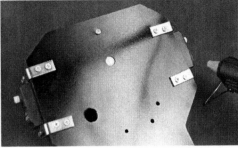

Figure 1-16. *RGB LED hot-glued*

As a final touch, we hot-glued a lens salvaged from an old headlamp on the top of the LED. This is not essential, but it will make the LED less blinding and better looking. If you have a lens, spread some hot glue on the bottom of the lens and stick it on top of the LED (Figure 1-17).

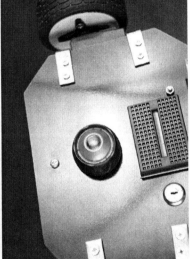

Figure 1-17. *Lens glued on top of the chassis*

Attaching the Power Switch to the Chassis

As we're going to be running our robot with a rechargeable battery, we'll need a way to turn it on and off. Any two-state switch will serve this purpose. Our choice was a key switch, but you can go with anything that pleases your eye.

Stick the switch in the hole and secure it with some hot glue (Figure 1-18). Some switches can be just screwed into the hole, making gluing unnecessary.

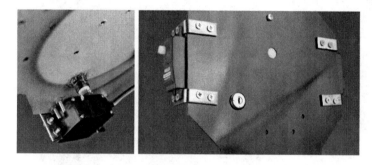

Figure 1-18. *Key switch hot-glued*

Attaching Arduino

Before attaching the Arduino to the robot, cover the bottom of the Arduino with a self-adhesive felt pad (Figure 1-19). This will prevent short circuits that could happen if the Arduino touched metal parts of the bot.

Figure 1-19. *Felt pad on the bottom and Arduino screwed in place.*

Use four 3x16 mm screws to secure the Arduino to the chassis (Figure 1-19).

Battery Holder

Cut 14 cm strip from both sides of Velcro. Tape hooks and loops sides together and push or drill a hole in the middle. Put a 3x10mm screw through the hole and mount the battery holder to the chassis (Figure 1-20).

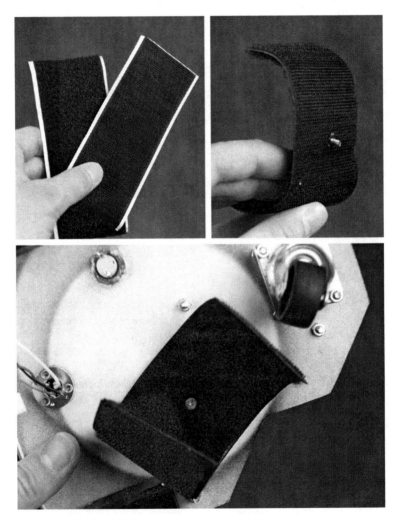

Figure 1-20. *14 cm strips of Velcro attached into the bottom of the chassis*

Attaching Solderless Breadboard

Remove the adhesive cover from the bottom of the breadboard and stick it in the center of the robot (Figure 1-21).

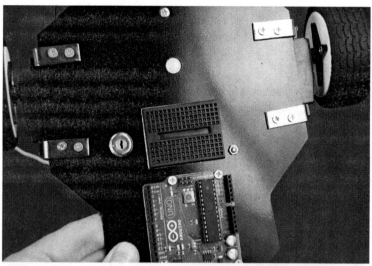

Figure 1-21. *Solderless breadboard in place; the mind-controlled robot is ready to be programmed*

ScrewShield Holds Wires in Place

Loose wires are probably the most annoying thing while building a prototype. Arduino's header pins can't hold the jumper wires securely. Especially when your robot is moving, wires pop out constantly. Luckily the ScrewShield pretty much solves this problem.

ScrewShield adds "wings" with terminal blocks to both sides of Arduino. Terminal blocks have screws so you can attach one or more wires firmly to any pin.

You have to do some soldering when building the ScrewShield, but it will pay off later on. As an added bonus, you'll get nice and relatively easy soldering practice. Of course you can make the robot we are building here without the ScrewShield, but we strongly suggest that you use it. The Appendix explains how to build the ScrewShield.

This completes your robot chassis! You now have a moving robot platform. In the next chapter, you'll be coding the mind-controlled robot part by part. What other projects could you do with your new robot chassis?

2/Coding

The code that controls your robot is simple. Well, it's simpler than one would expect for a sci-fi trick like mind control.

A simplified, pseudocode version of your robot would be as follows:

```
void loop()
{
        while (lineDetected()) turn();  // ❶
        forward(getAttention());        // ❷
}
```

❶ The reflectivity sensor tells if the robot is on top of a black line, which forms the border of the arena. When the robot is on the border, it turns until the sensor sees white again.

❷ Then the robot goes forward. The speed is read from the EEG headband.

Rinse and repeat.

In this coding part of the book, you'll write each part of the code that makes your robot tick.

Moving

Continuous rotation servos are motors you can easily control. With continuous rotation servos, you can only control the speed (forward, stop, back), but unlike standard servos, you can't set the servo to point at a specific angle.

There are three wires going to a servo: red (positive), black (ground), and white (data). To control a servo, you need to keep sending pulses to it. The length of the pulse tells how fast the full rotation servo should turn. The pulse length is very short, typically from 500 µs (microseconds) to 2000 µs. This means it can be as little as half a millisecond: 500 µs = 0.5 µs = 0.0005 seconds.

To keep the servo turning, you must send these pulses about every 20 ms, which is 50 times a second.

If you want a more through recap of how servos work, see the Insect Robot chapter in *Make: Arduino Bots and Gadgets* (MABG) from O'Reilly (2011).

Connect Servos

Let's start by connecting servos to Arduino. To keep the connectors in your servos intact, use a servo extension cable to connect to Arduino. Cut the male end of a servo extension cable. Strip wires from the end where you cut off the connector (Figure 2-1).

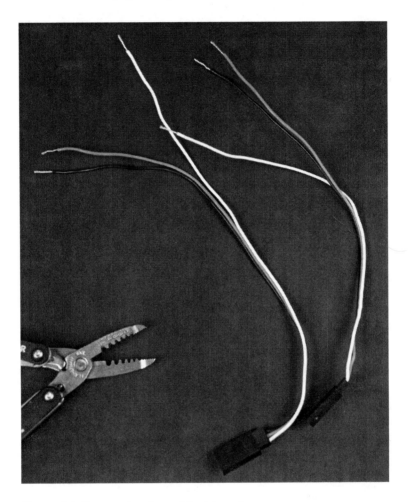

Figure 2-1. *Servo extension cable stripped*

Now connect the wires to Arduino. You can connect both the left and right servo right away. As shown in the circuit diagram (Figure 2-5), the red (positive) wire goes to Arduino +5V and black (ground) goes to GND (Figure 2-2). The white data wires go to Arduino data pins D2 (digital pin 2) and

Figure 2-2. *Ground and +5V connected to Arduino*

D3 (digital pin 3). Connect the left servo to digital pin D3 (Figure 2-3 and Figure 2-4).

 Every GND symbol means a connection to Arduino GND pin, which is also connected to the black 0V GND wire of the battery. This saves the trouble of having a circuit diagram full of wire going to ground.

Figure 2-3. *Connected to Arduino data pins; left servo goes to D3*

Figure 2-4. *Servos connected to extension cables*

Hello Servo

The Hello Servo program spins just one servo, so you know that your servos and connections are working. Figure 2-5 shows the connection diagram, even though here you'll use only one of the servos.

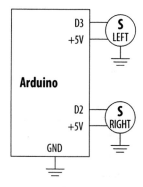

Figure 2-5. *Circuit diagram for forward.pde and helloservo.pde*

```
// helloservo.pde - Spin continuous rotation servo
// (c) Kimmo Karvinen & Tero Karvinen http://MindControl.BotBook.com

void setup()
{
        pinMode(3, OUTPUT);          // ❶
}

void loop()
{
        digitalWrite(3, HIGH);
        delayMicroseconds(2000);     // ❷
        digitalWrite(3, LOW);
        delay(10); // ms             // ❸
}
```

❶ You must always set the pin mode; otherwise it stays in an undefined state. OUTPUT means you can toggle the voltage of the pin: for example, HIGH or LOW for digital pins. INPUT means you can measure whether a pin is on or off.

❷ This is the length of the pulse, the time the pin is HIGH.

❸ It's a good idea to use a comment to specify the units at least once in your program. Millisecond is one thousand microseconds; i.e., 0.001 s = 1 ms = 1000 μs.

The servo will still turn if we have a longer wait, such as 20 ms. But using 10 ms will make it easy to count times in your head. For example, 100 iterations is 1000 ms = 1 s.

Calibrate Stopping Point

To calibrate the servos, tell them to stop. Then slowly adjust the calibration screw until they do.

To tell them to stop, simply change the pulse length in *helloservo.pde* to 1500 µs, which means stop:

```
delayMicroseconds(1500);
```

You can easily find this value and other specs from the data sheet. If you are using the Parallax continuous rotation servos, you can find the data sheet at *http://www.parallax.com/dl/docs/prod/motors/crservo.pdf*.

As you run the code, the servo will most likely crawl forward (because it is not yet calibrated). Locate the calibration screw on the side of the servo. Take your Phillips (crosshead) screwdriver and gently adjust the screw until the movement stops.

Calibrate the other servo by changing the pin in *helloservo.pde* to 2.

Full Speed Forward

Turn your robot into a racing car by running it forward, as fast as the servos can turn (Figure 2-5).

```
// forward.pde - Drive the robot forward
// (c) Kimmo Karvinen & Tero Karvinen http://MindControl.BotBook.com

const int servoLeftPin=3;
const int servoRightPin=2;

void setup()
{
        pinMode(servoLeftPin, OUTPUT);
        pinMode(servoRightPin, OUTPUT);
}

void loop()
{
        forward();           // ❶
}

void forward()
{
        for (int i = 0; i < 10; i++) {          // ❷
```

```
                    pulseServo(servoLeftPin, 1500+500);        // ❸
                    pulseServo(servoRightPin, 1500-500);       // ❹
          }
}

void pulseServo(int pin, int microseconds)
{
          digitalWrite(pin, HIGH);
          delayMicroseconds(microseconds);
          digitalWrite(pin, LOW);
          delay(5); // ms          // ❺
}
```

❶ Write functions with a clear purpose, so it's easy to reuse code.

❷ We'll do this 10 times. The loop starts i from zero, and keeps going as long as it's less than 10, which means it counts from 0 to 9. Each iteration of the loop takes 10 ms, so ten loops takes tenth of a second: 10 * 10 ms = 100 ms = 0.1 s.

❸ The compiler performs the math before generating the machine code, so writing 1500+500 is as fast as writing 2000. But the logic is much clearer when it's written out this way. If you wanted to be fancy, you could even create a new constant and write ServoStopped+500.

❹ The right servo is attached backwards to the robot frame. Therefore, to go forward, you must spin it the opposite direction compared to the left servo.

❺ Change the delay so that pulsing two servos takes 10 ms.

If your robot is not going straight, calibrate the stopping point by adjusting the built-in potentiometer with a screwdriver. If your robot is going backwards, swap the servo cables.

Other Ways to Control Servos

It's also possible to control servos with the built-in Arduino library, Servo.h. Servo.h comes with the Arduino IDE and provides an object-oriented interface to servos. It lets you control the servos by specifying degrees, like myservo.write(180). You could find it handy when you need to control many standard servos.

In the mind-controlled robot, we are using continuous rotation servos. They don't turn to specific positions defined by degrees, but instead they spin at different speeds. Using pulseServo(), the program flow is simple—and blocking: the function runs for a while, blocking all other activity, then returns. Servo.h uses a more advanced, interrupt-based flow, which does not block other activity.

Line Avoidance

To make your bot stay in the arena, you must teach it to avoid a black line. Then you can build an arena with big white paper as the floor and black tape as the border.

Connect the Reflection Sensor

Let's connect the reflection sensor to Arduino. Prepare the cable by cutting the end that doesn't fit to the sensor. In our case, the small white connector fit to the sensor and we left it in place. The big black connector didn't fit anywhere, so we cut it away. Strip the free wires for connecting to Arduino (Figure 2-6).

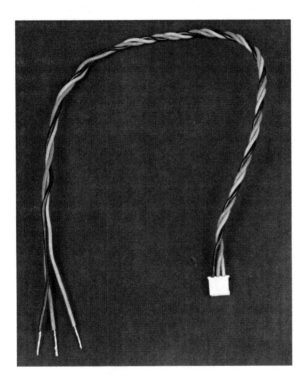

Figure 2-6. *Stripped sensor wire*

Connect free sensor wires to Arduino as shown in the circuit diagram for *helloreflection.pde* (Figure 2-7). Connect the red plus wire to +5V, and the black ground wire to GND. Connect the green data wire to D4. (Figure 2-8). Use the ScrewShield to keep the wires in place (Figure 2-9).

Figure 2-7. *Circuit diagram for helloreflection.pde*

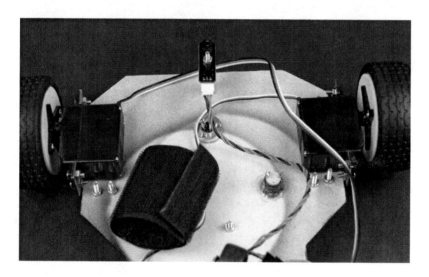

Figure 2-8. *Reflection sensor connected*

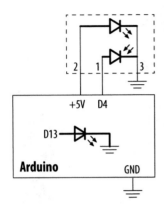

Figure 2-9. *Free wires connected to Arduino with ScrewShield*

Hello Reflection

For line avoidance, we use a typical reflectivity sensor. We read it with Arduino's `digitalRead()` function.

The value tells how much light is reflected back from the surface. It returns HIGH (true, 1) for a white surface. Black doesn't reflect light back, so the sensor gives a LOW value (false, 0).

```
// hellorefelection.pde - turn led on if line detected
// (c) Kimmo Karvinen & Tero Karvinen http://MindControl.BotBook.com

const int tinyLedPin=13;        // ❶
const int linePin=4;

void setup()
{
        pinMode(tinyLedPin, OUTPUT);
        pinMode(linePin, INPUT);
}

void loop()
{
        if (LOW == digitalRead(linePin)) {        // ❷
                digitalWrite(tinyLedPin, HIGH);
        } else {
                digitalWrite(tinyLedPin, LOW);
        }
        delay(20);        // ❸
}
```

❶ We use the tiny surface-mounted LED that's built into Arduino to show if we've detected a line. The reflectivity sensor also has its own LED, so don't confuse them (Arduino's built-in LED is connected to pin 13 and is right on the Arduino board itself).

❷ LOW from the reflectivity sensor means black.

❸ Long running loops should always have some delay. Otherwise the program would take 100% of CPU time on any single core CPU (such as the one on Arduino), no matter how fast it was.

Test your reflectivity sensor by holding it near white and black surfaces.

Don't Cross the Black Line

Lighting an LED is a nice way to test the sensor, but what we really want to do is turn to avoid the border. Something like this:

```
void loop()
{
        while (lineDetected()) turn();
        forward();
}
```

You already have all the knowledge to do this (in separate parts) from the previous examples, so let's combine them. Connect the servos and reflection sensor according to the circuit diagram for *avoidline.pde* (Figure 2-10).

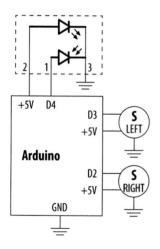

Figure 2-10. *Circuit diagram for avoidline.pde*

```
// avoidline.pde - turn when black line seen
// (c) Tero Karvinen & Kimmo Karvinen http://MindControl.BotBook.com

const int linePin=4;
const int servoLeftPin=3;
const int servoRightPin=2;

void setup()
{
        pinMode(linePin, INPUT);
        pinMode(servoLeftPin, OUTPUT);
        pinMode(servoRightPin, OUTPUT);
}

void loop()
{
        while (lineDetected()) turn();        // ❶
        forward();
}

bool lineDetected()
{
        return ! digitalRead(linePin);        // ❷
}

void forward()
{
        for (int i = 0; i < 10; i++) {
                pulseServo(servoLeftPin, 1500+500);
                pulseServo(servoRightPin, 1500-500);
        }
}

void turn()        // ❸
{
        for (int i = 0; i < 30; i++) { // ❹
                pulseServo(servoLeftPin, 1500+500);
                pulseServo(servoRightPin, 1500+500); // ❺
        }
}

void pulseServo(int pin, int microseconds)
{
        digitalWrite(pin, HIGH);
        delayMicroseconds(microseconds);
        digitalWrite(pin, LOW);
        delay(5);
}
```

❶ Keep turning as long as the sensor sees a black border. When the `while()` loop is done, the bot is away from the border. Because `lineDe tected()` already returns true or false, we don't have to put in an explicit comparison. Because there is just a single line (`turn()`) inside the while-loop, we don't need braces (`{}`) here.

❷ When the bot is in the white arena, linePin is HIGH, which Arduino represents internally as the number 1. In a boolean context, 1 is interpreted as true (and 0 is false). And according to boolean algebra, not true is false. Thus, `lineDetected()` returns false when the robot is on a white surface.

❸ The new function `turn()` turns right. It's almost same as `forward()`, but the count of loop iterations and the direction of right servo have changed.

❹ Every call to `turn()` takes some time by running `pulseServo()` in a loop. Otherwise the bot would only turn just enough to go parallel to the black border. Feel free to play with the values to alter the turns to work as you need them to.

❺ Make the right wheel spin the opposite direction. Just change the plus to minus. We send the same signal, 2000 μs, to both servos, because the right one is mounted backwards on the frame.

Battery, No Strings Attached

Your line-avoiding robot is otherwise great, but it can be a bit of an annoyance to run after it while you're holding the USB cable. It's time for your robot to get its very own power source.

We can't use a regular 9-volt battery for the bot, as it doesn't have enough juice to power motors, the LED, and the NeuroSky dongle. It's also much nicer to use a high-capacity rechargeable battery, because it can keep your robot running 10 times longer without the need for a battery change.

Choosing Rechargeable Batteries

We used DualSky 1300 mAh 30C 2s1p 7.4 V battery (Figure 2-11). What do all these values mean? 1300 mAh (milliamp hours) or 1.3 Ah is the battery capacity. Higher value means longer running time. 30C is the discharge capacity for one hour. In this case 1300 mAh * 30 = 39000 mAh = 39 Ah. 2s1p means that battery has two 3.7 V cells in parallel. So together they give you 7.4 V, the output current. The recommended input voltage for Arduino Uno is 7–12 V.

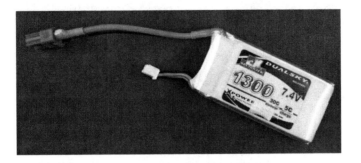

Figure 2-11. *Rechargeable lithium-polymer battery pack*

 Always use strong batteries carefully. Acquire and follow battery-specific instructions. A short circuit can lead to heating, fire, or even explosion.

Connecting the Battery and Power Key

Batteries often use one set of cables for charging and other for discharge (powering your device). In our battery, the blue connector is for discharge and small white one is for charging (Figure 2-11). Be careful to use the correct one.

Let's connect the battery and the key switch. The ground wire from the battery connects directly to Arduino, but the positive wire has a power switch in the middle (Figure 2-12). You need to use a connector for the battery, so that you can disconnect it for charging.

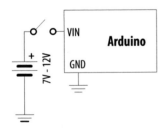

Figure 2-12. *Circuit diagram of battery and power switch*

The battery cable has a connector on one end and free wires on the other. It connects the battery to the power switch and to Arduino GND. Use a connector suitable for your battery. Our battery came with a male DC3 connector, so we use a DC3 female for the battery cable.

Leave the black free wire 25 cm long. Cut one red free wire to 8 cm and another one to 15 cm. Strip the wires and solder the black wire and shorter red wire to the switch.

Put heat shrink tubing over the soldered joints (Figure 2-13). Leave some tubing for the other end of the red wire, too (Figure 2-14). You can't add tube after you've soldered both ends.

Solder red wires to the power switch and put some heat shrink tubing on them (Figure 2-15). Connect the free red wire to Arduino VIN and the black wire to Arduino GND (Figure 2-16).

Now you have connected the battery and the switch (Figure 2-17).

The power source is often omitted from circuit diagrams. We didn't use a symbol for USB power earlier, so we won't be using one for the battery all of the time either.

Well done! You have now built a complete line-avoiding robot. Let it play in the arena for a while and enjoy a cup of coffee (or appropriate celebratory beverage).

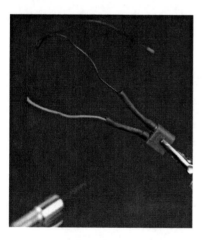

Figure 2-13. *Putting heat shrink tubing on the cable for connecting battery to Arduino and power switch*

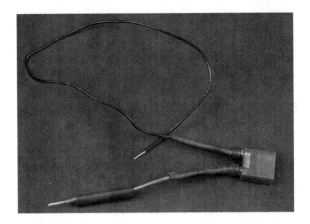

Figure 2-14. *Adding a piece of heat shrink tubing before soldering to the power switch*

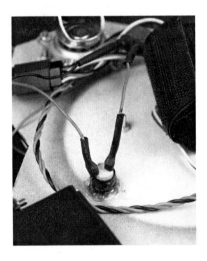

Figure 2-15. *Positive wires soldered to the power switch*

Figure 2-16. *Switch wires connected to the ScrewShield*

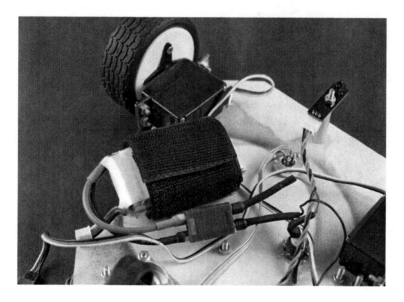

Figure 2-17. *Battery and power switch connected*

Bells and Whistles

LEDs and piezo speakers create user interface for many embedded devices. Even though they are low-key compared to computer displays and big speakers, they serve an important function. Users expect to get feedback on what they are doing.

Using LEDs and speakers to describe program state also helps coding. It would be difficult to keep a moving robot continuously on a USB leash. Sounds and lights on the robot will tell you what's happening, so you don't have to divide your attention between the serial monitor and the device. And of course, you don't have to fight with a lost serial connection.

Red, Green, and Blue LED

An RGB LED has three LEDs in one package (Figure 2-18). The colors of those LEDs are the primary colors: red, green, and blue (Figure 2-19). When you mix colors by turning on multiple LEDs, you get secondary colors such as cyan, magenta, and yellow.

Figure 2-18. *An RGB LED has three LEDs in one package*

Figure 2-19. *RGB LED primary colors*

The RGB LED you use in this project has one positive leg and three negative legs. This configuration is called *common anode*. The positive wire is always connected to Arduino's +5V.

When your data pins (D9, D10, D11) are at +5V, there is no voltage difference between the LED's legs. All the LEDs are off.

When you want to light a LED, you set the pin to a low voltage. You can do this with analogWrite(0). Now the data pin has 0V and common anode has +5V, and the LED is lit.

Soldering RGB LED

Find out which pin is which by applying power to the LED and testing it. One pin will be plus. When you connect GND to another pin, you get one of the primary colors: red, green, or blue. Use scotch tape or a marker to mark R, G, B, and +.

You can use a battery to test your LED. If you use Arduino +5V and GND as power source, it would be wise to put a small resistor in series with the LED. A suitable resistor would be 470 Ohm (third stripe brown) or 1 kOhm (third stripe red). That said, we foolhardily connected our LED without resistors and it didn't burn out. In the actual robot, the LED is used with PWM (pulse width modulation), which means that the voltage doesn't stay at its peak voltage for long. Still, it would be a good idea to use a resistor in the bot, too.

Solder the wires in place (Figure 2-20). Any jumper wire will do, but we used ribbon cable. Choose colors sensibly: green and blue LEDs can have green and blue wires. We reserved the red color for +5V and used another one for the red LED.

Figure 2-20. *Wires soldered to RGB LED*

Connect the wires as shown in Figure 2-21. Common anode (common plus) to Arduino +5V, and LEDs red to D9, green to D10, and blue to D11.

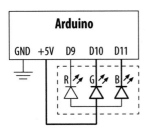

Figure 2-21. *Circuit diagram for hellorgb.png*

Hello RGB

```
// hellorgb.pde - Mix colors with RGB LED
// (c) Kimmo Karvinen & Tero Karvinen http://MindControl.BotBook.com

const int redPin=9;         // ❶
const int greenPin=10;
const int bluePin=11;

void setup()
{
        pinMode(redPin, OUTPUT); // ❷
        pinMode(greenPin, OUTPUT);
        pinMode(bluePin, OUTPUT);
}
```

```
void loop()
{
        setColor(255, 0, 0);        // ❸
        delay(1000); // ms          // ❹

        setColor(255, 255, 255);
        delay(1000);
}

void setColor(int red, int green, int blue)
{
        analogWrite(redPin, 255-red);       // ❺
        analogWrite(greenPin, 255-green);
        analogWrite(bluePin, 255-blue);
}
```

❶ analogWrite() uses digital pins. All Arduinos support AnalogWrite on D9, D10, and D11; most also support D3, D5, and D6.

❷ Many analogWrite() examples don't set +pinMode(pin, OUTPUT)+— even the ones in the Arduino reference. But we still prefer to do it.

❸ This is a simple function to set the color. A web color like #ff00ff would be setColor(255, 0, 255) which is red and blue: indigo.

❹ Colors set stay there until you set them to something else. You should wait a while for the user to see them before you change colors again.

❺ Writing a low value (0) to red pin lights it with maximum brightness. That's why you must calculate the difference of maximum value (255) and the amount of color you want.

Did your LED alter between red and white? Very good, every LED in your RGB LED is working.

You can try setting your LED to primary colors red, green and blue. If you can remember the additive color model from high school, you can also get secondary colors cyan, magenta, and yellow. Or search Wikipedia for "RGB color model".

Beeping Piezo

"Beep, beep!" Piezo beepers are everywhere: in cash registers, cameras— even your PC speaker is one. A stream of on-off pulses makes a piezo speaker vibrate, making air vibrate and thus making a sound. If you do it fast enough you get a note. Our favorite note, A4, comes with 440 pulses a second (440 Hz).

Push the piezo speaker into the breadboard on top of your robot. Piezos have polarity: their positive side is usually marked with a plus (Figure 2-22). Connect the positive side to Arduino digital pin D12, using a jumper wire with the data wire color, like yellow, green, or blue. Connect the negative side to Arduino GND with black jumper wire (Figure 2-23 and Figure 2-24).

Figure 2-22. *Small plus marks the positive leg of piezo speaker*

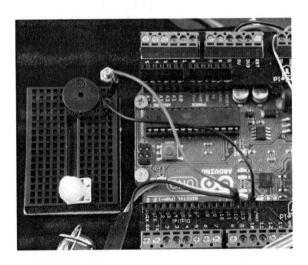

Figure 2-23. *Piezo speaker connected to Arduino with a breadboard*

Figure 2-24. *Circuit diagram for hellospeaker.pde*

```
// hellospeaker.pde - Beep a piezo speaker
// (c) Kimmo Karvinen & Tero Karvinen http://MindControl.BotBook.com

const int speakerPin=12;

void setup()
{
        pinMode(speakerPin, OUTPUT);
}

void loop()
{
        wave(speakerPin, 440, 500);        // ❶
        delay(50);          // ❷
        wave(speakerPin, 540, 500);
        delay(50);
        wave(speakerPin, 300, 400);
        delay(1000);
}

void wave(int pin, float frequency, int duration)
{
        float period = 1/frequency * 1000 * 1000; // microseconds // ❸
        long int startTime = millis();
        while (millis()-startTime < duration) {        // ❹
                digitalWrite(pin, HIGH);
                delayMicroseconds(period / 2);        // ❺
                digitalWrite(pin, LOW);
                delayMicroseconds(period / 2);
        }
}
```

❶ This **wave()** function is all you really need to grasp to play some tones. In addition to pin number, you can tell it the frequency (A4 is 440 Hz) and duration in milliseconds.

❷ A short delay between notes makes them sound clearer.

❸ You can treat this **wave()** function like a black box and just call it. But if you insist, we'll show you what's inside. Period T, the duration of a single wave (one HIGH + one LOW) is 1/frequency (T=1/f).

❹ This uses the duration that was passed into the function. This code uses a common idiom to do something for a duration: store the starting time, check how much time we've spent since then, and see if it's more than the duration.

❺ You can create a square wave by setting a pin half of the time LOW and half of the time HIGH. The code that calls the function can specify the frequency, which is the count of HIGHs and LOWs per second (1Hz = 1/s).

Why not use **tone()**? The Arduino IDE ships with **tone()** and **noTone()**. We had minor timing problems with **tone()**. Though our **wave()** is definitely not perfect either, it's very short and is simpler than **tone()**. This makes problems much easier to debug. Also, **wave()** helps you understand how a piezo speaker actually works.

Setting Threshold with a Potentiometer

A potentiometer (pot) is a resistor you can adjust.

Your complete robot will use a potentiometer to set the threshold, which is the minimum level of attention required to move the robot.

Connect the center leg of the pot to Arduino analog pin A0. Connect either one of the side pins of the pot to Arduino GND (Figure 2-25 and Figure 2-26).

```
// hellopot.pde - Print values from a potentiometer.
// (c) Kimmo Karvinen & Tero Karvinen http://MindControl.BotBook.com

const int potPin=A0;        // ❶

void setup()
{
        pinMode(potPin, INPUT);      // ❷
        digitalWrite(potPin, HIGH);  // internal pullup // ❸

        Serial.begin(9600); // bit/s // ❹
}

void loop()
{
        int n=analogRead(potPin); // ❺
        Serial.println(n);        // ❻
        delay(400);               // ❼
}
```

Figure 2-25. *Potentiometer on the breadboard*

❶ analogRead() only works with analog pins (A0, A1...). The pin name has the letter A in it; e.g., analogRead(A0).

❷ Many examples (including Examples→Analog→AnalogInput) don't set pinMode(). But you should. For example, setting the internal pullup on the next line is only guaranteed to work when the pins are in INPUT mode.

❸ Set the internal pullup resistor for the pin. This connects the pin to +5V through a large 20 kOhm resistor. When the pin is not connected to ground, it goes to +5V. This is absolutely required, and enabling it in code saves you from having to physically add a resistor to your circuit.

❹ Open the serial port. We had some trouble with the serial monitor, so we used a slower speed than the maximum 115,200 bits per second.

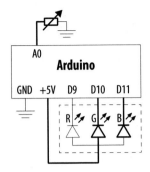

Figure 2-26. *Circuit diagram for hellopot.pde*

❺ `analogRead()` returns a number between 0 (0V) and 1023 (5V).

❻ Our potentiometer doesn't necessarily have the optimal range of resistance. We display the values on the serial monitor so we can observe the minimum and maximum values.

❼ We wait almost half a second to avoid flooding the serial buffer.

Make note of minimum and maximum values you can get from your potentiometer. The one we used got values from 14 to 236. Usually we want to make sure that the user can easily turn the knob to min and max. That's why it's good to also look at smallest and largest values that stay visible without fluctuation. For our potentiometer, these were about 13 and 231.

You might run into some problems with the serial monitor. It could manifest itself as the Arduino IDE freezing, IDE menus freezing, serial monitor not opening, or upload not working. If this happens, disconnect the USB and restart the Arduino IDE.

 If you mistakenly uploaded code that floods the serial buffer when it runs, don't worry. Disconnect Arduino and restart the IDE. Compile the Blink example that comes with Arduino. Connect the USB and quickly upload Blink. You might need to try this a couple of times to get fast enough (and you can try hitting the reset button just before you press upload).

Everything But Your Mind

You have the parts now; let's put them together. This code will have everything but mind control. Here's what will happen:

Start the bot, and it plays a sound. You can set a movement threshold with the potentiometer. `getAttention()` is essentially a dummy function that always returns 50% attention. When you turn the threshold under this, the bot starts moving. The multicolor LED shows the threshold level, from 0% blue to 100% red.

When you stop the bot by turning the threshold above 50%, it will slowly grind to a halt. It's not really that heavy, but the code is faking inertia. Fake inertia will make the robot move smoothly with mind control.

Connect the parts (Figure 2-27) and upload the code.

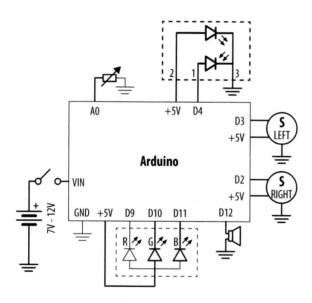

Figure 2-27. *Circuit diagram for allbutmind.pde*

Code Structure

This code uses only one global variable, speed. For fake inertia to work, we need to know the last speed the robot was moving at as we slow down. This global variable is used only in `loop()` to keep the program structure clean. The speed variable is sent to functions as a regular parameter.

The three input sensors are processed with lineDetected(), getAtten tion(), and getThreshold().

Outputs include servos, LEDs, and speakers. The most important output functions are forward(), turn(), sayReady(), and setBlueToRed(). The rest of the output functions are just helper functions to outputs: pulseServo(), wave(), and setColor().

```
// allbutmind.pde - a mind controlled robot without mind control
// (c) Tero Karvinen & Kimmo Karvinen http://MindControl.BotBook.com


const int linePin=4;
const int servoLeftPin=3;
const int servoRightPin=2;
const int potPin=A0;
const int redPin=9;
const int greenPin=10;
const int bluePin=11;
const float potMin=14.0-1;        // ❶
const float potMax=236.0-5;
const int speakerPin=12;

float speed=0.0;          // ❷

void setup()
{
        pinMode(linePin, INPUT);
        pinMode(servoLeftPin, OUTPUT);
        pinMode(servoRightPin, OUTPUT);
        pinMode(potPin, INPUT);
        digitalWrite(potPin, HIGH); // internal pullup
        pinMode(speakerPin, OUTPUT);

        sayReady();        // ❸
}

void loop()
{
        while (lineDetected()) turn();

        float at=getAttention();        // ❹
        float tr=getThreshold();        // ❺
        speed=fakeInertia(speed, at, tr);        // ❻

        forward(speed);
}

float fakeInertia(float speed, float at, float tr)
{
        if (at<tr)        // ❼
```

```
                                speed-=0.05;
                else
                                speed=at;                // ❽
                if (speed<0) speed=0.0;                   // ❾
                return speed;
}

/*** Inputs ***/

bool lineDetected()
{
                return ! digitalRead(linePin);
}

float getAttention()
{
                return 0.5; // dummy value for testing            // ❿
}

float getThreshold()
{
                int x=analogRead(potPin);            // ⓫
                float tr=(x-potMin)/potMax;          // ⓬
                setBlueToRed(tr);
                return tr;
}

/*** Outputs ***/

void forward(float speed)
{
                for (int i = 0; i < 10; i++) {
                                pulseServo(servoLeftPin, 1500+500*speed);
                                pulseServo(servoRightPin, 1500-500*speed);
                }
}

void turn()
{
                for (int i = 0; i < 30; i++) {
                                pulseServo(servoLeftPin, 1500+500);
                                pulseServo(servoRightPin, 1500+500);
                }
}

void pulseServo(int pin, int microseconds)
{
                digitalWrite(pin, HIGH);
                delayMicroseconds(microseconds);
                digitalWrite(pin, LOW);
                delay(5);
```

```
        }

void setBlueToRed(float redPercent)
{
        int red=redPercent*255;          // ⓭
        int blue=(1-redPercent)*255;
        setColor(red, 0, blue);
}

void setColor(int red, int green, int blue)
{
        analogWrite(redPin, 255-red);
        analogWrite(greenPin, 255-green);
        analogWrite(bluePin, 255-blue);
}

void sayReady()
{
        wave(speakerPin, 440, 40);
        delay(25);
        wave(speakerPin, 300, 20);
        wave(speakerPin, 540, 40);
        delay(25);
        wave(speakerPin, 440, 20);
        wave(speakerPin, 640, 40);
        delay(25);
        wave(speakerPin, 540, 40);
        delay(25);
}

void wave(int pin, float frequency, int duration)
{
        float period = 1/frequency * 1000 * 1000; // microseconds
        long int startTime = millis();
        while (millis()-startTime < duration) {
                digitalWrite(pin, HIGH);
                delayMicroseconds(period / 2);
                digitalWrite(pin, LOW);
                delayMicroseconds(period / 2);
        }
}
```

❶ This line and the next specify limits of the potentiometer you measured
earlier with *hellopot.pde*. We've shortened the range a bit, so that the user
can easily reach the minimum and maximum value. Even though 14 and
236 could be represented as integers, they must be specified as *float*
here. That's because we use the numbers in division, and we want the
results to be floating points. With integers, 1/2=0. But usually, we really
want a float, such as 1.0/2.0=0.5.

❷ This is a global variable, which could be read and modified from any function. But we don't do that! We only use this in `loop()`, and pass it to functions as a parameter.

❸ Play a sound.

❹ Get the attention level from NeuroSky MindWave. Currently, the dummy function always returns 0.50, meaning 50% attention.

❺ Threshold is read from the potentiometer.

❻ When attention falls below threshold, we want to stop slowly. Note how we pass speed as a normal parameter, so that `fakeInertia()` doesn't have to access the global variable. All `speed`, attention `at`, and threshold `tr` are percentages (such as 63%), represented by floating point numbers (such as 0.63).

❼ At low concentration, slow down to a halt.

❽ If attention is above threshold, use attention as speed. So maximum attention gets maximum speed.

❾ Don't go backwards.

❿ We'll implement this later. The attention will be read from NeuroSky MindWave headband.

⓫ `analogRead()` returns a value from 0 (0V) to 1023 (5V). But this raw value isn't really convenient to play with.

⓬ We convert the raw value to percent of the max value possible with our potentiometer. For example, turning it to minimum results in x=14, making tr equal to (14-13)/231.0 = 1/231.0 = 0.0.

⓭ We show percentage as a color. Zero is blue, one is red. Anything in between is a mixture.

Almost there! After this code, you only have to handle reading attention to finish your mind-reading robot.

Measuring Your Brains with MindWave

Now let's take the next step and make the robot respond to your brain activity!

Hack MindWave Dongle

To connect the MindWave dongle directly to Arduino, we need to modify it a little bit. See Figure 2-28.

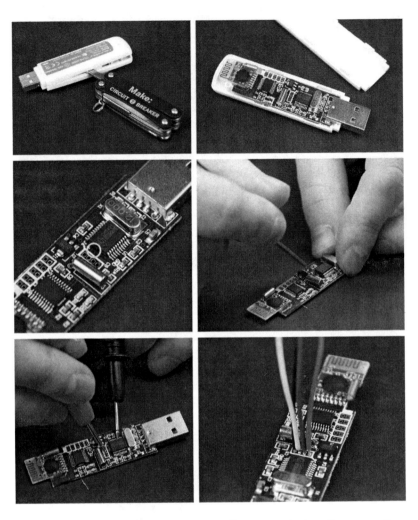

Figure 2-28. *Hacking the dongle: remove the cover; destroy the connection marked with green circle; confirm you've broken the connection with multimeter; solder wires to dongle*

Start by removing the dongle cover by lifting it from the seam with a screwdriver.

Use a razor knife to destroy the marked connections. The point of doing this is to disconnect the dongle USB port to enable you to read data from the RX and TX pins.

Use a multimeter with a conductivity setting to confirm that there is no connection between the chip pins 3 and 4 and the four connection pins shown in Figure 2-28.

Cut four different colored 8 cm wires. Solder the wires to the dongle as shown in Figure 2-29. Start with the red wire for the first pin, marked with "+" on the board. Solder the green wire to the dongle's TX pin. Solder the white wire to the dongle's RX pin. Finally, solder the black wire to the dongle's GND. If you use a flux pen, it will make the process easier, because the old solder joints will flow more easily.

Figure 2-29. *Dongle connections*

Level Conversion with Resistors

The NeuroSky USB dongle uses 3.3 volts to represent HIGH. Arduino uses 5 volts, 50% more. To avoid breaking the dongle, we must convert the level of voltage.

Most level converters are integrated circuits, black chips with circuits inside. Unfortunately, we never have the right chip at hand when we need it. In this project, you'll build a level converter with just two resistors!

Three of the four dongle pins don't need anything but jumper wire: +3.3V, GND, and TX. The resistors are needed for the dongle RX pin, so that it doesn't get the full 5V from Arduino's TX.

When connecting any two circuits, we want comparable voltage levels. When the ground level 0V is the same in both circuits, all voltage levels are comparable. Connect the dongle's black GND wire to Arduino's GND.

The dongle also needs power. Connect the red positive wire to Arduino's 3.3 volt output ("3V3").

Connect the dongle's TX directly to Arduino's RX (pin 0). When dongle TX (transmission) is sending the bit 0 (LOW), the voltage is 0V, same as Arduino LOW.

What about when the dongle's TX sends bit 1 to Arduino RX? Surprisingly, no resistors are needed here. When dongle TX is sending the bit 1 (HIGH), the voltage is 3.3V. This is one-third less than Arduino's 5V HIGH. But it's still over half of 5V! As 3.3V is more than 2.5V (Arduino's threshold), Arduino considers this HIGH. Because the voltage is lower than expected, nothing will get too hot.

Voltage Divider

Arduino's TX sending 5V to a dongle pin made for 3.3V requires more tricks.

Connect Arduino TX (pin 1) with a 1.8 kOhm resistor to Dongle TX. Connect Dongle TX to GND with a 3.3 kOhm resistor.

Arduino sending bit 0 (LOW) is the simpler case. In the top of Figure 2-30, you can see dongle TX connected to 0V through both resistors. Obviously, it ends up in ground level, 0V.

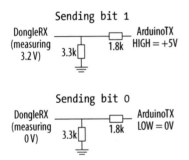

Figure 2-30. *Level converter with just two resistors*

What about Arduino sending bit 1, HIGH? Here, the two resistors act as a voltage divider.

On the top, there is +5V. Then there are two resistors, and finally 0V. Between the two resistors, dongle RX measures voltage. According to Ohm's law (U=IR), the voltage gradually reduces to 0V as you follow the resistors.

The total resistance is 1.8 kOhm + 3.3 kOhm = 5 kOhm. If you start from +5V and measure the voltage at 1.8 kOhm, the voltage is reduced by 36% (1.8 kOhm / 5 kOhm).

So 64% of 5 V voltage is left. It seems we've chosen the correct resistors, as 64% of 5 V is 3.2 V. This is almost the same as the target, dongle HIGH level 3.3 V.

Now you know the "two resistor level converter" trick.

We also had to see if the hacked dongle would work without a level converter. In our foolhardy tests, we noticed that NeuroSky dongle could take 5 volts— at least for a short while. However, we prefer to keep our components in reliable, stable condition. That's why we normally use the level converter.

Hello Attention!

Connect your hacked dongle to Arduino. Connect the wires using the breadboard in your bot (Figure 2-31) according to the circuit diagram (Figure 2-32).

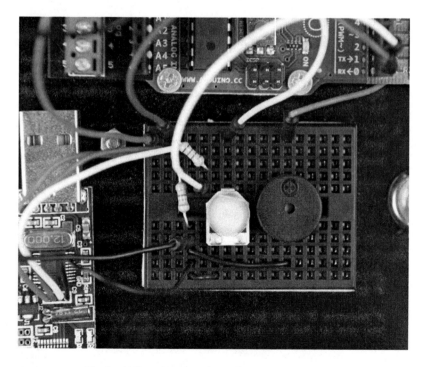

Figure 2-31. *Hacked NeuroSky dongle connected to breadboard*

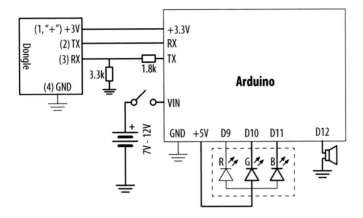

Figure 2-32. *Circuit diagram for helloattention.pde*

Upload helloattention.pde and try it out.

Try controlling the LED with your mind:

1. Turn off both the NeuroSky EEG headband and Arduino.
2. Turn on Arduino. The Main LED turns green.
3. Turn on the NeuroSky EEG headband. Tiny LEDs in the NeuroSky headband and dongle are red. When the wireless connection is working, the tiny LEDs turn blue. When the robot can get information from the headband, the main LED turns yellow.
4. Place the headband on your head. When the robot gets the signal, the main LED shows your attention. The main LED will be a color from blue to red.

As you focus, the main LED will turn red. When your mind wanders and attention goes down, the main LED will turn blue. In between, it's indigo, a mixture of blue and red.

Got helloattention working? Great! Now you can start building your own attention-controlled projects.

```
// helloattention.pde - Show attention level (EEG) with LED color.
// (c) Kimmo Karvinen & Tero Karvinen http://MindControl.BotBook.com

/* Disconnect TX and RX jump wires from Arduino when uploading from IDE.
Turn robot on, then in a couple of seconds turn headband on. */

const int redPin = 9;
const int greenPin = 10;
const int bluePin = 11;
```

```
const int tinyLedPin = 13;
const int speakerPin = 12;

int tinyLedState = HIGH;

void setup()
{
        pinMode(redPin, OUTPUT);
        pinMode(greenPin, OUTPUT);
        pinMode(bluePin, OUTPUT);
        pinMode(tinyLedPin, OUTPUT);
        pinMode(speakerPin, OUTPUT);

        Serial.begin(115200); // bit/s   // ❶
        connectHeadset();                // ❷
}

void loop()
{
        float att = getAttention();      // ❸
        if (att > 0)         // ❹
                setBlueToRed(att);
        toggleTinyLed();     // ❺
}

/*** Headset ***/

void connectHeadset()
{
        setGreen();
        delay(3000);
        Serial.write(0xc2);      // ❻
        setWhite();
}

byte readOneByte()
{
        while (!Serial.available()) {    // ❼
                delay(5); // ms
        };
        return Serial.read();
}

float getAttention()
{       // return attention percent (0.0 to 1.0)
        // negative (-1, -2...) for error
        byte generatedChecksum = 0;      // ❽
        byte checksum = 0;
        int payloadLength = 0;
        byte payloadData[64] = {
                0
```

```
};
int poorQuality = 0;
float attention = 0;

Serial.flush(); // prevent serial buffer from filling up // ❾

/* Sync */
if (170 != readOneByte()) return -1;          // ❿
if (170 != readOneByte()) return -1;

/* Length */
payloadLength = readOneByte();
if (payloadLength > 169) return -2;           // ⓫

/* Checksum */
generatedChecksum = 0;
for (int i = 0; i < payloadLength; i++) {      // ⓬
        // Read payload into array:
        payloadData[i] = readOneByte();
        generatedChecksum += payloadData[i];
}
generatedChecksum = 255 - generatedChecksum;
checksum = readOneByte();
if (checksum != generatedChecksum) return -3;   // ⓭

/* Payload */
for (int i = 0; i < payloadLength; i++) {       // ⓮
        switch (payloadData[i]) {
        case 0xD0:
                sayHeadsetConnected();
                break;
        case 4:        // ⓯
                i++;        // ⓰
                attention = payloadData[i]; // ⓱
                break;
        case 2:
                i++;
                poorQuality = payloadData[i];
                if (200 == poorQuality) {
                        setYellow();            // ⓲
                        return -4;
                }
                break;
        case 0xD1: // Headset Not Found
        case 0xD2: // Headset Disconnected
        case 0xD3: // Request Denied
        case -70:
                wave(speakerPin, 900, 500);
                setWhite();
                return -5;
                break;
```

```
                case 0x80:          // skip RAW          // ⑲
                        i = i + 3;
                        break;
                case 0x83:          // skip ASIC_EEG_POWER
                        i = i + 25;
                        break;
            } // switch
        } // for

        return (float)attention / 100;        // ⑳
}

/*** Outputs ***/

void setBlueToRed(float redPercent)
{
        int red = redPercent * 255;
        int blue = (1 - redPercent) * 255;
        setColor(red, 0, blue);
}

void setGreen()
{
        setColor(0, 255, 0);
}

void setYellow()
{
        setColor(255, 255, 0);
}

void setWhite()
{
        setColor(100, 100, 100);
}

void sayHeadsetConnected()
{
        wave(speakerPin, 440, 40);
        delay(25);
        wave(speakerPin, 300, 20);
        wave(speakerPin, 540, 40);
        delay(25);
        wave(speakerPin, 440, 20);
        wave(speakerPin, 640, 40);
        delay(25);
        wave(speakerPin, 540, 40);
        delay(25);
}

void setColor(int red, int green, int blue)
```

```
{
        analogWrite(redPin, 255 - red);
        analogWrite(greenPin, 255 - green);
        analogWrite(bluePin, 255 - blue);
}

void toggleTinyLed()
{
        tinyLedState = !tinyLedState;
        digitalWrite(tinyLedPin, tinyLedState);
}

void wave(int pin, float frequency, int duration)
{
        float period = 1 / frequency * 1000 * 1000; // microseconds
        long int startTime = millis();
        while (millis() - startTime < duration) {
                digitalWrite(pin, HIGH);
                delayMicroseconds(period / 2);
                digitalWrite(pin, LOW);
                delayMicroseconds(period / 2);
        }
}
```

❶ Communication with the dongle is very fast, 115.2 kbit/s. It's always a good idea to specify the units in your comments.

❷ At very beginning, we set the main LED to green to ask the user to turn on the headset.

❸ getAttention() is the heart of the program. It returns attention as a percentage, from 0.0 (0%) to 1.0 (100%).

❹ We only show attention with the main LED if there is some. Otherwise, we leave the main LED set to whatever color it is currently set to: green, white, or yellow. Function getAttention() could have set it to one of the following: green (turn on headband), white (no headband-dongle-arduino connection), or yellow (no EEG, even though headband-dongle-arduino connection works).

❺ We blink Arduino's surface mounted LED (pin 13) on every iteration of loop() to confirm that we're still running. Normally, the blinking is so fast it looks like flicker or almost as though the LED was continuously on.

❻ 0xc2 is instruction for the dongle to try connecting to any headband it can find. This has to happen briefly after the headband is turned on, because the headband is trying to get a connection, too.

❼ Wait until a byte comes from serial. If this means forever, then so be it. But readOneByte() is guaranteed to return a byte. As a loop with no delay takes infinite capacity of any single core CPU, we add a very short wait in the loop.

❽ Variables are local, so we don't have to care about those when outside `getAttention()`.

❾ If the serial buffer fills, Arduino seems to crash. With `flush()`, we ignore all data currently in the serial buffer. Losing this data is not critical, so long as we call `getAttention()` often.

❿ Return an error if we don't see the sync sequence 170 170. The `return` statement ends this function. This is much better than putting the meat of the function inside multiple +if+s.

⓫ The maximum length is from the NeuroSky reference documentation.

⓬ Both calculate the checksum and read payload into the array `payload Data[]`.

⓭ Compare calculated checksum to the number that came with the packet. The checksum calculation algorithm is from NeuroSky documentation.

⓮ Iterate through each byte in the `payloadData[]` array. Here, `i` is the number of payload bytes we have processed.

⓯ In this case, the payload indicated that it's of the *field* type (4 means attention).

⓰ We increment the counter `i` here, too. It would be more common to increment `i` in just the loop header `for(...; i++)`. But in the NeuroSky protocol, different data types have different lengths, so incrementing `i` here gives the simplest solution.

⓱ As it's obvious from here that we're reading attention, no comment about the meaning of field type 4 is needed.

⓲ This indicates a problem to the user by changing the main LED color. It will be left in place while the problem persists, because `loop()` only changes color if `attention > 0`.

⓳ Even if long values are not interesting to us, we have to skip them. A comment is necessary, because the meaning is not obvious from the code.

⓴ Floating point numbers from 0.0 to 1.0 are practical for percentages. For example, you can multiply by a floating point percentage value. We use an explicit cast to float to ensure we get a floating point value.

Now that it works, would you like to know how? Or if it doesn't work yet, would you like to know the details so you can fix it?

As of Arduino 1.0, which was still in testing at the time of this writing, the behavior of flush() is changing significantly. If you are using Arduino 1.0, check the book's catalog page (see "How to Contact Us" on page xiv) for new example code, or visit the authors' site at *http://botbook.com*.

NeuroSky Protocol

How do you begin when faced with an unknown protocol? We did some Googling, and found that the NeuroSky reference documentation and more was available on the NeuroSky website. For example, see "MindWave and Arduino" (NeuroSky 2011), "MindSet Communications Protocol" (NeuroSky 2010), and "arduino_tutorial" in the NeuroSky wiki (*http://developer.neuro sky.com/docs/doku.php?id=start*).

The NeuroSky dongle communicates over the serial port. We have wires going to Arduino RX and TX.

Every packet is of this format: sync (170 170), payload length, payload, and a checksum (see Table 2-1).

Table 2-1. *NeuroSky packet*

Meaning	Example	Comment
Sync	170 170	Always the same
Payload length	4	Most common length in this project
Payload	4 86	4 attention, value 86/100 = 86%
Payload cont.	2 0	2 poor signal, value 0 signal OK
Checksum	163	255−(sum of each payload byte)

Even though the reference documentation saved us the tedious work of reverse-engineering the protocol, there were surprises around the corner. Coping with serial buffer overflow, long delays caused by moving parts, and doing voltage level changes without additional circuits were problems to solve.

NeuroSky, the manufacturer of MindWave, provides excellent documentation. See "MindSet Communications Protocol" and "MindWave Dongle Communications Protocol."

Some code samples on the Web use everyday decimal format (170), but some use hexadecimal (0xaa or just AA). We've compiled a list of the most common code numbers (Table 2-2).

You can also convert numbers with Python. Just start Python at the command line and you'll get an interactive console where you can type things like *hex(170)* or *0xaa*. Or you can use *http://www.wolframalpha.com* or Google. Most of the NeuroSky protocol control characters we use are outside ASCII. The last ASCII printable character is 126 tilde ~; the last ASCII character is 127 DEL.

Table 2-2. *Hex numbers used in helloattention.pde*

Hex	Dec	Used in
-0x46	-70	Fail
0x80	128	Skip 2 byte RAW Wave value
0x83	131	Skip 24 byte ASIC_EEG_POWER value
0xAA	170	Sync
0xC2	194	Initiate connection
0xD0	208	Headset connected
0xD1	209	Headset not found
0xD2	210	Headset disconnected
0xD3	211	Request denied

Complete Mind-Controlled Robot

Now that we have every part working and tested, it should be easy to put it together, right? Wrong.

Even though reading constant stream of packets from the dongle works with LEDs, it's a different case for turning the servos. LEDs, being semiconductors, take negligible time to change state. Servos move things that have mass, and are very slow compared to LEDs.

The previous attention routine assumed it will be called at suitable intervals. Here, we have to make an attention-reading function that can both be called very often and survive slight pauses.

Build the final circuit (Figure 2-33) and upload mindcontrol.pde.

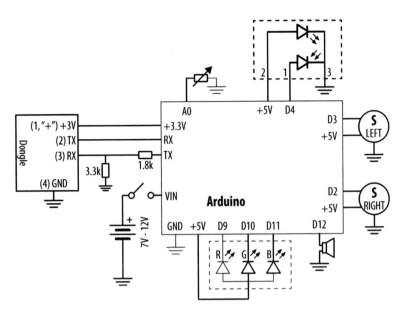

Figure 2-33. *Circuit diagram for complete robot, mindcontrol.pde*

```
// helloattention.pde - Show attention level (EEG) with LED color.
// (c) Kimmo Karvinen & Tero Karvinen http://MindControl.BotBook.com

/* Disconnect TX and RX jump wires from Arduino when
   uploading from IDE.
   Turn robot on, then in a couple of seconds turn headband on.
 */

const int redPin = 9;
const int greenPin = 10;
const int bluePin = 11;
const int tinyLedPin = 13;
const int speakerPin = 12;
const int servoRightPin = 2;
const int servoLeftPin = 3;

const int linePin = 4;
const int potPin = A0;

const float potMin = 14.0 - 1;
const float potMax = 236.0 - 5;

int tinyLedState = HIGH;
```

```
// percent variables, 0.0 to 1.0:
float tr = -1.0;
float attention = 0;
float speed = 0.0;

void setup()
{
        pinMode(redPin, OUTPUT);
        pinMode(greenPin, OUTPUT);
        pinMode(bluePin, OUTPUT);
        pinMode(tinyLedPin, OUTPUT);
        pinMode(speakerPin, OUTPUT);

        pinMode(servoRightPin, OUTPUT);
        pinMode(servoLeftPin, OUTPUT);

        pinMode(linePin, INPUT);

        pinMode(potPin, INPUT);
        digitalWrite(potPin, HIGH);          // internal pullup

        Serial.begin(115200);        // bit/s
        connectHeadset();
}

void loop()
{
        while (lineDetected()) turn();
        updateAttention();          // ❶
        tr = getThreshold();        // ❷
        if (attention > 0)
                setBlueToRed(attention);
        if (attention > tr) {
                speed = attention;          // ❸
        } else {
                speed = speed * 0.98;          // ❹
        }
        forward(speed);          // ❺
        toggleTinyLed();
}

float getThreshold()
{
        int x = analogRead(potPin);
        return (x - potMin) / potMax;
}

/*** Input: Other ***/

bool lineDetected()
{
```

```
                  return !digitalRead(linePin);
}

/*** Input: Headset ***/

void connectHeadset()
{
        setGreen();
        delay(3000);
        Serial.write(0xc2);
        attention = 0;
        setWhite();
}

byte ReadOneByte()
{
        while (!Serial.available()) { }
        return Serial.read();
}

float updateAttention()
{
        byte generatedChecksum = 0;
        byte checksum = 0;
        int payloadLength = 0;
        byte payloadData[64] = { 0 };
        int poorQuality = 0;

        while ((170 != ReadOneByte()) && (0 < Serial.available())) {
                smartFlush();          // ❻

        }
        if (170 != ReadOneByte()) return -1;          // ❼

        /* Length */
        payloadLength = ReadOneByte();
        if (payloadLength > 169)
                return -2;

        /* Checksum */
        generatedChecksum = 0;
        for (int i = 0; i < payloadLength; i++) {
                payloadData[i] = ReadOneByte(); // Read payload into array
                generatedChecksum += payloadData[i];
        }
        generatedChecksum = 255 - generatedChecksum;
        checksum = ReadOneByte();
        if (checksum != generatedChecksum) {
                return -3;
        }
```

```
        /* Analyse payload */
        for (int i = 0; i < payloadLength; i++) {
                switch (payloadData[i]) {
                case 0xD0:
                        sayHeadsetConnected();
                        break;
                case 4:
                        i++;
                        attention = payloadData[i] / 100.0; // ❽
                        break;
                case 2:
                        i++;
                        poorQuality = payloadData[i];
                        if (170 < poorQuality) {          // max 200
                                setYellow();
                                attention = 0.0;
                                return -4;
                        }
                        break;
                case 0xD1:      // Headset Not Found
                case 0xD2:      // Headset Disconnected
                case 0xD3:      // Request Denied
                case -70:
                        wave(speakerPin, 900, 500);
                        attention = 0.0;
                        setWhite();
                        return -5;
                        break;
                case 0x80:
                        i = i + 3;
                        break;
                case 0x83:
                        i = i + 25;
                        break;
                } // switch
        } // for
}

void smartFlush()
{
        if (128 / 2 < Serial.available()) {   // buffer is 128 B // ❾
                Serial.flush();
        }
}

/*** Outputs: Servos ***/

void forward(float speed)
{
        if (speed <= 0)
                return;
```

```
                for (int i = 0; i < 2; i++) {
                        pulseServo(servoLeftPin, 1500 + 500 * speed);
                        pulseServo(servoRightPin, 1500 - 500 * speed);
                }
        }

        void turn()
        {
                for (int i = 0; i < 20; i++) {
                        pulseServo(servoLeftPin, 1500 + 500);
                        pulseServo(servoRightPin, 1500 + 500);
                }
        }

        void pulseServo(int pin, int microseconds)
        {
                digitalWrite(pin, HIGH);
                delayMicroseconds(microseconds);
                digitalWrite(pin, LOW);
                updateAttention();        // ❿
        }

        /*** Outputs: LED ***/

        void setBlueToRed(float redPercent)
        {
                int red = redPercent * 255;
                int blue = (1 - redPercent) * 255;
                setColor(red, 0, blue);
        }

        void setGreen()
        {
                setColor(0, 255, 0);
        }

        void setYellow()
        {
                setColor(255, 255, 0);
        }

        void setWhite()
        {
                setColor(100, 100, 100);
        }

        void setColor(int red, int green, int blue)
        {
                analogWrite(redPin, 255 - red);
                analogWrite(greenPin, 255 - green);
                analogWrite(bluePin, 255 - blue);
```

```
        }

        void toggleTinyLed()
        {
                tinyLedState = !tinyLedState;
                digitalWrite(tinyLedPin, tinyLedState);
        }

        /*** Outputs: Speaker ***/

        void sayHeadsetConnected()
        {
                wave(speakerPin, 440, 40);
                delay(25);
                wave(speakerPin, 300, 20);
                wave(speakerPin, 540, 40);
                delay(25);
                wave(speakerPin, 440, 20);
                wave(speakerPin, 640, 40);
                delay(25);
                wave(speakerPin, 540, 40);
                delay(25);
        }

        void wave(int pin, float frequency, int duration)
        {
                float period = 1 / frequency * 1000 * 1000; // microseconds
                long int startTime = millis();
                while (millis() - startTime < duration) {
                        digitalWrite(pin, HIGH);
                        delayMicroseconds(period / 2);
                        digitalWrite(pin, LOW);
                        delayMicroseconds(period / 2);
                }
        }
```

❶ updateAttention() now updates the global variable **attention**. This makes it possible to call updateAttention() in any part of our code, such as when sending pulses to servos. It doesn't return a value like getAttention() in helloattention.

❷ Attention threshold is the minimum attention where the bot moves. This makes it much clearer to test subjects that they can start and stop moving with the power of their minds. Threshold is set with the small potentiometer.

❸ If there is enough attention, that will be our speed. 50% attention means 50% of maximum speed.

❹ Attention is too low, so slow down as if the robot were heavy. It loses 2% of speed in every iteration of **loop()**. Feel free to try other values.

❺ Moving forward uses a floating point percentage (0.0 to 1.0), the same range as attention and threshold. There is no hurry returning from forward, because we call `updateAttention()` from `pulseServo()`, the low-level command called by `forward()`.

❻ Skip until we get to the beginning of NeuroSky packet, marked by sync sequence start 170. The function `smartFlush()` only flushes the buffer if it's near overflow.

❼ The second byte of sync sequence 170 170.

❽ Update global variable attention. This is the main purpose of `updateAttention()`.

❾ The buffer is 128 bytes. We can't let it overflow, as that seems to crash the bot. But we want to catch up to whatever messages came while we were busy moving servos. So we only flush (delete the contents of) the buffer if it's half full. The 50% fill value was found by experimentation.

❿ We can move servos for two minutes if we want to. This is very useful if you want to add preprogrammed sequences, such as long turns or more obstacle avoidance. The attention value will be updated and the serial buffer kept away from overflow. Feel free to put `updateAttention()` in any of your own functions.

Congratulations! You've now finished your mind-controlled robot!

This is not the end, this is the beginning. What's the next mind-controlled project in your mind?

See you at *http://BotBook.com*!

Figure 2-34. *Finished mind-controlled robot*

A/Building the ScrewShield

Start by putting two 8-pin headers (see Figure A-1) into the pin holes of the "digital" wing (the one with RX and TX pins). Use some poster putty such as Blu-Tack to hold pins in place (Figure A-2). Otherwise they will easily get misaligned when you are soldering them from the other side.

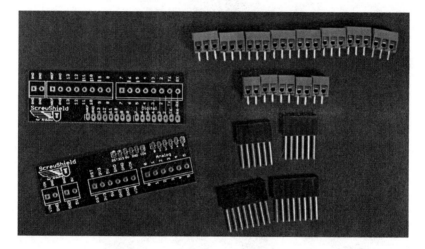

Figure A-1. *Contents of the ScrewShield package*

Mount the wing to third-hand tool or similar (Figure A-3). You can't hold it yourself while you are soldering, as you need one hand for soldering iron and other for the solder. Turn the wing upside down and solder the pins in place.

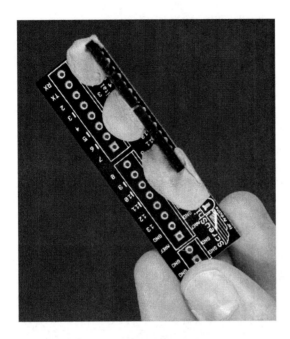

Figure A-2. *Poster putty holds two 8-pin headers in place*

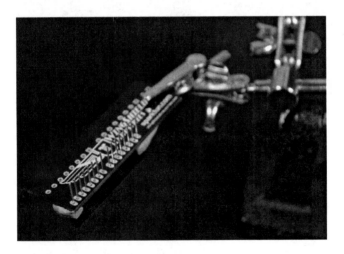

Figure A-3. *Headers ready to be soldered*

Now you need to do the same thing to the other wing. Secure the 6-pin headers with poster putty. Turn the wing upside down and solder (Figure A-4).

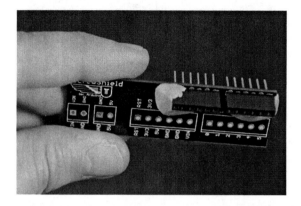

Figure A-4. *"Analog" wing with 6-pin headers*

The ScrewShield package includes a variety of terminal blocks. They can be connected by vertically sliding them together. Form two 6-pin blocks and two 8-pin blocks (Figure A-5).

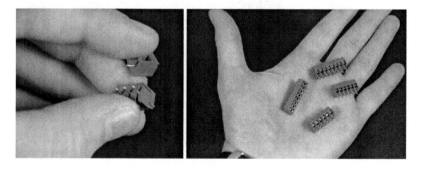

Figure A-5. *Forming two 6-pin blocks and two 8-pin blocks*

Now use poster putty to put the terminal blocks in place. 6-pins blocks go to the "analog" wing and 8-pins blocks to the "digital" wing (Figure A-6).

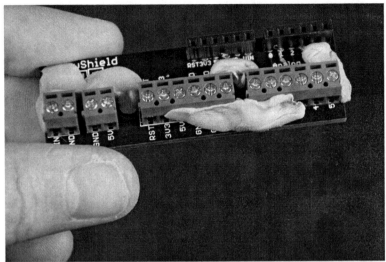

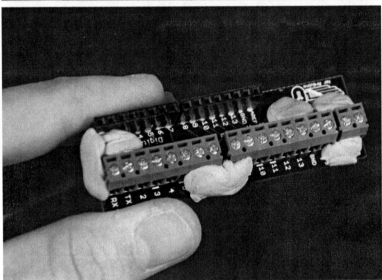

Figure A-6. *"Analog" and "Digital" wings with the terminal blocks*

Solder the terminal blocks to wings. Push the wings to Arduino ports, and you are done (Figure A-7).

Figure A-7. *ScrewShield attached to Arduino*

About the Authors

Tero Karvinen teaches Linux and embedded systems in Haaga-Helia University of Applied Sciences, where his work has also included curriculum development and research in wireless networking. He previously worked as CEO of a small advertisement agency. Tero's education includes a master's of science in economics.

Kimmo Karvinen works as CTO at a hardware manufacturer that specializes in integrated AV and security systems. Before that, he worked as a marketing communications project leader and as a creative director and partner at an advertisement agency. Kimmo's education includes a masters of art.

Get even more for your money.

Have it your way.

CPSIA information can be obtained at www.ICGtesting.com
Printed in the USA
BVOW010614131212

308042BV00011BA/307/P